Barthélemy BAWAR

Report on the water quality monitoring network in 2023

Barthélemy BAWAR

Report on the water quality monitoring network in 2023

National network for monitoring water quality by the General Directorate for Water Resources laboratory: year 2023

ScienciaScripts

Imprint

Cover image: www.ingimage.com

This book is a translation from the original published under ISBN 978-620-6-71137-7.

Publisher:
Sciencia Scripts
is a trademark of
Dodo Books Indian Ocean Ltd. and OmniScriptum S.R.L publishing group

120 High Road, East Finchley, London, N2 9ED, United Kingdom
Str. Armeneasca 28/1, office 1, Chisinau MD-2012, Republic of Moldova, Europe
Printed at: see last page
ISBN: 978-620-7-61706-7

FOREWORD

This report highlights the crucial importance of monitoring water quality in Burkina Faso, given the many sources of pollution, such as industrial discharges, small-scale mining and the intensive use of chemical fertilisers in agriculture. These factors compromise the physico-chemical quality of the water, making it unsuitable for a variety of uses. A national water quality monitoring network, initially consisting of 32 points, has been in place since 1992, but has proved inadequate in the face of increasing pollution. In 2018, an expansion of the network to 80 sites was undertaken, but security constraints have prevented some points from being monitored in 2023. The report highlights the difficulties associated with collecting, processing and using data, while identifying the needs to ensure adequate monitoring of raw water quality at both national and cross-border levels.

CONTENTS

INTRODUCTION

Water quality is an important parameter that affects all aspects of the well-being of ecosystems and human beings, such as the health of populations, food production, economic activities and biodiversity. In Burkina Faso, the quality of raw water raises many questions due to uncontrolled industrial discharges, small-scale exploitation of mining resources and intensive use of chemical fertilisers in agriculture. These various sources of pollution contribute to the deterioration of the physico-chemical quality of the water, making it unsuitable for the desired uses. Monitoring water quality has become an imperative for the protection of the environment and the health of living beings. For this reason, it is important to monitor water quality and its evolution over time and space. To achieve this, a national network of 32 water quality monitoring points spread across the four (04) river basins has been in place since 1992.This non-optimal network, with a very low density, could not meet the needs for monitoring water quality in a context of increasing natural and man-made pollution of water resources. The Directorate General for Water Resources (DGRE) therefore increased the number of sampling points in 2018.The current raw water quality monitoring network, managed by the Water Quality Department (SQE), comprises 80 measurement sites, which are sampling points at the location of certain piezometric sites and hydrometric stations.For this year 2023, around thirty points could not be monitored due to the security situation in certain areas of the country.This report therefore highlights the difficulties in the system for collecting, processing and using data, and defines the requirements for adequate monitoring in order to perpetuate the monitoring of raw water quality.

I. PRESENTATION OF THE WATER QUALITY MONITORING NETWORK

The DGRE's current water quality monitoring network is a nationwide network consisting mainly of 80 sites, including 36 hydrometric stations and 44 piezometric sites and boreholes, as shown in **Figure 1** :

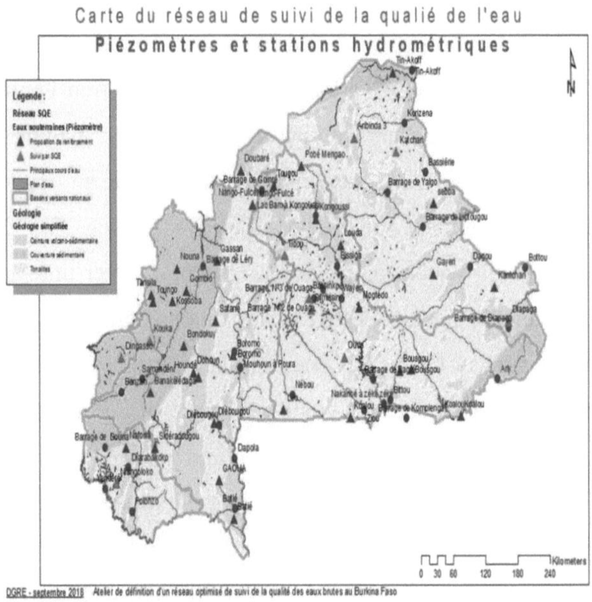

Figure 1: Map of water quality monitoring network

II. PRESENTATION OF RESULTS

II.1. Sources of data

In 2023, three (03) water sampling campaigns were carried out on the water quality monitoring network, the first in May/June, the second in September/October and the third in December. The water samples taken were analysed in situ and at the DGRE laboratory.

II.2. Strengths and weaknesses of data

II.2.1. Forces

- The first strength of these data is that they come from samples taken and analysed by the same laboratory: the DGRE laboratory, which uses standardised sampling and analysis methods;

- The second strength of these data lies in the fact that they form part of a space of fairly short time: all the points were sampled in 2023 ;

- The third is that there were three sampling campaigns over the course of the year. 2023 at the same points in the monitoring network.

II.2.2. Weaknesses

The data reveal many weaknesses, the most important of which are highlighted below:

- The first weakness of the data is that it is not representative. Of the 80 water points planned for the country, only around thirty were sampled and analysed. This was due to insecurity in certain sampling zones.

- The second weakness relates to the network, which conceals disparities in water quality underground ;

- The third weakness concerns the absence of certain important parameters such as **cyanide, mercury, lead, cadmium, chromium, arsenic and pesticide residues** in all the data (surface water and groundwater), despite the fact that some areas contain mining sites and companies. These parameters are very important for groundwater quality. This is due to a breakdown observed since 2022 in the atomic absorption spectrometer, particularly the oven mode, which would enable these parameters to be measured. parameters. It should be pointed out that this device has been in need of maintenance since it was installed some ten years ago, which is hampering its operation.

- Surface water data also lacks pollution parameters such as **Biochemical Oxygen Demand (BOD5), Chemical Oxygen Demand (COD) and other organic and mineral micro-pollutants.** This is due to the lack of equipment for measuring BOD5 and COD chemical indices at the DGRE laboratory.

Tables (I, II, III, IV) summarise the physico-chemical measurements and analyses carried out at the sites in the water quality monitoring network in 2023.

Table I: Analysis results for the groundwater sites studied

Name of water points	Sampling period	T°C	PH	Conductivity (µS/cm)	Turbidity (NTU)	Calcium (mg/L)	Magnesium (mg/L)	Bicarbonate (mg/L)	Chloride (mg/L)	Nitrates (mg/L)
Piezometer Niangoloko	May/June	31,4	5,79	99,6	10,61	4,64	0,048	58,56	0,77	1,93
	September/October	30	6,38	115,9	9,5	5,6	1,92	59,78	0,1	0,5
	December	30,3	5,22	92,8	2,68	5,4	1,9	58,11	0,1	0,4
	Average values	30,5	5,79	102,76	7,596	5,21	1,28	58,81	0,32	0,94
Piezometer Nafona F1	May/June	29	7,47	209	22,17	34,24	2,49	123,95	2,83	6,22
	September/October	27,9	6,55	126,3	23,98	52,96	0,57	161,04	2,7	5,1
	December	29,2	7,58	104,1	23,5	30,86	0,48	155,06	2,5	4,7
	Average values	28,7	7,2	146,46	23,21667	39,35	1,18	146,68	2,676	5,34
Nafona F2 piezometer	May/June	29,4	6,88	163,5	8,74	29,76	1,72	123,7	1,43	4,17
	September/October	28	6,8	126,4	26,76	31,28	11,8	201,42	1,2	4,9
	December	29,9	7,53	106,2	2,98	29,26	9,81	154,45	0,8	3,5
	Average values	29,1	7,07	132,03	12,82	30,1	7,77	159,85	1,14	4,19
Gaoua piezometer	May/June	30,1	6,01	210	0,2	21,92	5,04	152,5	0,24	0,46
	September/October	27,6	6,2	227	0,73	20,56	2,16	139,202	0,3	0,5
	December	30,6	6,39	340	0,49	23,5	3,1	143,22	1,3	1,5
	Average values	29,43	6,2	259	0,47	21,99	3,43	144,97	0,61	0,82
Boromo piezometer	May/June	30,4	6,92	454	70,01	39,76	11,088	254,492	1,5	0,23
	September/October	31,8	6,43	524	98,47	46	6,768	220,82	24,9	0,6

Name of water points	Sampling period	T°C	PH	Conductivity (µS/cm)	Turbidity (NTU)	Calcium (mg/L)	Magnesium (mg/L)	Bicarbonate (mg/L)	Chloride (mg/L)	Nitrates (mg/L)
	December	29,8	6,88	489	18,34	38	10,78	250,89	23,8	0,6
	Average values	30,67	6,74	489	62,27	41,25	9,54	242,06	16,73	0,47
Drilling of Navielgane	May/June	31,1	6,26	197	0,61	20,48	2,832	142,74	0,23	0,1
	September/October	30,7	6,16	195,4	0,56	16,8	7,53	120,78	0,2	0,1
	December	30,7	6,2	196	0,66	17,1	7,54	118,9	0,2	0,1
	Average values	30,83	6,20	196,1	0,61	18,12	5,96	127,47	0,21	0,1
Mogtedo piezometer	May/June	31	7,9	423	38,88	81,28	5,18	295,24	1,16	1,86
	September/October	30,2	7,04	432	31,19	75,92	7,63	294,75	0,9	4,1
	December	25,8	6,38	492	4,46	95,22	5,45	293,12	0,5	3,8
	Average values	29	7,10	449	24,84	84,14	6,087	294,37	0,85	3,25
Silmissin piezometer	May/June	32,2	7,53	230	0,7	20,8	9,216	139,08	85,18	0,25
	September/October	31,4	5,82	221	37	19,28	9,312	146,034	0,6	3,8
	December	28,5	6,1	168	5,5	15,8	8,32	136,9	0,5	3,6
	Average values	30,7	6,48	206,33	14,4	18,62	8,94	140,67	28,76	2,55
Louda piezometer	May/June	32	8,1	371	30,46	11,6	21,84	313,54	1,67	0,11
	September/October	32	6,89	279	33,9	34,4	29,28	283,65	1,1	0,1
	December	32,4	6,79	392	1,23	40,98	22,86	291,54	2,05	0,1
	Average values	32,1	7,26	347,3	21,86	28,99	24,66	296,243	1,60	0,1
Bassinko piezometer	May/June	31	8,5	162,3	3,77	19,36	4,08	101,26	55,19	2,01
	September/October	32,2	7,58	169,97	5,78	17,52	8,68	111,14	1,6	3,3

Name of water points	Sampling period	T°C	PH	Conductivity (µS/cm)	Turbidity (NTU)	Calcium (mg/L)	Magnesium (mg/L)	Bicarbonate (mg/L)	Chloride (mg/L)	Nitrates (mg/L)
	December	27,7	7,55	271	11,68	21,12	6,76	115,19	1,6	2,3
	Average values	30,3	7,87	201,09	7,07	19,33	6,50	109,196	19,46333	2,53
Tibou piezo meter	May/June	31,7	6,94	204	0,35	16,48	5,808	125,904	0,82	4,85
	September/October	29,1	7,28	308	6,05	26,13	5,52	130,784	1	5
	December	29,5	6,59	294	1,69	19,28	3,024	127,64	0,8	4,5
	Average values	30,1	6,93	268,6	2,697	20,63	4,784	128,109	0,87	4,78
Léo piezo meter	May/June	30,9	6,27	251	3,38	22,8	11,568	179,46	0,42	0,14
	September/October	30,7	6,7	212,6	0,6	16	7,53	138,95	0,4	0,1
	December	29,1	6,3	201	8,47	13	7,12	135,97	0,3	0,05
	Average values	30,2	6,42	221,53	4,15	17,26	8,73	151,46	0,37	0,09
Bobo-Dioulasso P14	May/June	29,8	5,44	56,2	15	5,21	2,5	65,16	32,11	6,29
	September/October	31,3	5,83	201	67,42	12,8	0,336	64,66	30,21	7,09
	December	30,9	5,16	82,4	402	6,87	2,16	61,55	27,71	5,79
	Average values	30,6	5,47	113,2	161,47	8,29	1,66	63,79	30,01	6,39
Kari borehole	May/June	31,5	7,02	519	9,7	62,88	21,35	326,9	2,5	20,5
	September/October	30,8	6,85	312	13,73	43,38	20,4	325,12	2,7	17,6
	December	31,5	7,02	292	11,7	45,89	23,52	320,96	2,1	19,12
	Average values	31,2	6,96	374,3	11,71	50,71	21,75	324,32	2,43	19,07

Table II: Analysis results for groundwater sites (Continued)

Name of water points	Sampling period	Phosphates (mg/L)	Sulphates (mg/L)	Copper (mg/L)	Zinc (mg/L)	Manganese (mg/L)	Potassium (mg/L)	Sodium (mg/L)
Niangoloko piezometer	May/June	0,55	0,23	< 0,05	0,078	0,89	2,15	13,68
	September/October	0,9	0,2	< 0,05	< 0,05	0,06	1,51	13,1
	December	0,2	0,2	< 0,05	< 0,05	0,04	0,91	10,05
	Average values	0,55	0,21	< 0,05		0,33	1,52	12,27
Nafona F1 piezometer	May/June	0,65	6,8	0,111	0,083	0,086	1,972	3,908
	September/October	0,8	9,3	< 0,05	< 0,05	0,06	< 0,5	< 0,5
	December	0,7	8,6	< 0,05	< 0,05	< 0,05	1,83	5,69
	Average values	0,7	8,23					
Nafona F2 piezometer	May/June	0,53	6,21	< 0,05	0,056	0,108	2,306	9,461
	September/October	0,6	10,3	< 0,05	< 0,05	< 0,05	2,72	18,19
	December	0,5	9,3	< 0,05	< 0,05	< 0,05	1,77	15,9
	Average values	0,54	8,60				2,265	14,5
Gaoua piezometer	May/June	1,44	0,1	< 0,05	< 0,05	0,055	1,38	21,67
	September/October	1,9	0,4	< 0,05	< 0,05	< 0,05	1,3	20,17
	December	2,9	1,4	0,05	0,05	0,055	2,3	23,1
	Average values	2,08	0,63				1,66	21,64
Boromo piezometer	May/June	0,33	1,26	< 0,05	< 0,05	0,09	3,6	10,3
	September/October	1,3	1,2	< 0,05	< 0,05	0,18	12,5	15,15
	December	1,3	1,2	< 0,05	< 0,05	0,19	10,5	16,19
	Average values	0,97	1,22			0,15	8,86	13,88

Name of water points	Sampling period	Phosphates (mg/L)	Sulphates (mg/L)	Copper (mg/L)	Zinc (mg/L)	Manganese (mg/L)	Potassium (mg/L)	Sodium (mg/L)
Navielgane borehole	May/June	1,6	2,17	< 0,05	< 0,05	0,044	2,644	18,33
	September/October	1,8	2,4	< 0,05	< 0,05	< 0,05	2,58	16,4
	December	1,7	2,5	< 0,05	< 0,05	< 0,05	1,98	19,4
	Average values	1,7	2,35				2,40	18,04
Mogtedo piezometer	May/June	1,78	0,92	< 0,05	< 0,05	0,093	2,24	11,13
	September/October	2,3	1,8	< 0,05	< 0,05	0,098	1,67	11,03
	December	2,1	1,5	< 0,05	< 0,05	< 0,05	0,97	10,81
	Average values	2,06	1,40				1,62	10,99
Silmissin piezometer	May/June	1	8,91	< 0,05	< 0,05	0,052	4,782	13,8
	September/October	1,4	1,6	< 0,05	< 0,05	0,07	4,5	12,58
	December	1,3	1,5	< 0,05	< 0,05	0,05	4,12	11,55
	Average values	1,23	4,			0,05	4,46	12,64
Louda piezometer	May/June	1,85	3,9	< 0,05	0,051	1,365	4,103	22,43
	September/October	2,1	0,6	< 0,05	< 0,05	1,03	3,53	18,08
	December	2,1	0,9	< 0,05	< 0,05	1,9	3,9	17,9
	Average values	2,01	1,8			1,43	3,84	19,47
Bassinko piezometer	May/June	0,58	16,95	< 0,05	< 0,05	0,048	2,393	9,431
	September/October	0,8	1	< 0,05	< 0,05	< 0,05	2,1	9,07
	December	0,7	1,08	< 0,05	< 0,05	< 0,05	2,1	9,7
	Average values	0,69	6,34				2,19	9,40
Tibou piezometer	May/June	0,23	2,37	< 0,05	< 0,05	0,07	0,8	17,3

Name of water points	Sampling period	Phosphates (mg/L)	Sulphates (mg/L)	Copper (mg/L)	Zinc (mg/L)	Manganese (mg/L)	Potassium (mg/L)	Sodium (mg/L)
	September/October	0,8	2,7	< 0,05	< 0,05	0,05	1,88	29,31
	December	0,56	1,97	< 0,05	< 0,05	0,05	2,87	28,06
	Average values	0,53	2,34			0,05	1,85	24,89
Léo piezometer	May/June	0,9	1,92	< 0,05	< 0,05	0,14	14,5	14,4
	September/October	1,4	0,5	< 0,05	< 0,05	0,12	4,96	14,26
	December	0,4	0,5	< 0,05	< 0,05	0,11	4,65	13,98
	Average values	0,9	0,97			0,12	8,03	14,21
Bobo Dioulasso P14	May/June	0,4	2,47	0,05	2,1	0,36	6,05	19,07
	September/October	0,5	3,17	0,05	2,2	0,41	5,21	15,11
	December	0,46	2,92	0,05	2,4	0,38	7,94	13,08
	Average values	0,45	2,85	0,05	2,23	0,38	6,4	15,75
Kari borehole	May/June	2,1	3,1	< 0,05	< 0,05	0,05	0,85	24,08
	September/October	1,9	2,8	< 0,05	< 0,05	0,06	0,57	25,48
	December	1,5	2,5	< 0,05	< 0,05	0,05	0,83	26,45
	Average values	1,83	2,8			0,05	0,75	25,33

Table III: Analysis results for the surface water sites studied

Name of water points	Sampling period	T°C	pH	Conductivity (µS/cm)	Turbidity (NTU)	Calcium (mg/L)	Magnesium (mg/L)	Bicarbonate (mg/L)	Chloride (mg/L)
Douna Barrage	May/June	28,7	7,11	31,6	37,25	2,88	1,92	21,106	0,5
	September/October	29,9	6,6	25,7	37,48	2,68	0,048	24,4	0,7
	December	28,5	7,15	31,4	37,62	3,12	0,92	23,11	0,7
	Average values	29,03	6,95	29,56	37,45	2,89	0,96	22,87	0,63
Diaraba koko	May/June	27,9	6,63	48,4	444,2	4,88	0,912	21,96	2,4
	September/October	26,5	6,72	64,9	21,7	11,7	3,16	42,7	5,86
	December	27,8	6,6	97,9	69,21	10,87	2,91	38,95	5,41
	Average values	27,4	6,65	70,4	178,37	9,15	2,32	34,53	4,55
Bougouriba	May/June	27,4	6,85	38,3	617,5	18,32	20,496	258,27	0,9
	September/October	29,4	6,93	87,6	37,85	4,48	1,968	39,04	1,48
	December	27,5	6,81	56,7	109,1	7,32	3,42	58,24	2,55
	Average values	28,1	6,86	60,86	254,81	10,04	8,628	118,51	1,64
Mouhoun Boromo	May/June	26,5	6,75	39,7	1058	2,4	2,832	14,152	0,6
	September/October	31,1	6,93	70	159	17,12	8,35	104,68	1,48
	December	24,3	6,77	127,1	111,2	12,48	7,64	101,16	0,98
	Average values	27,3	6,81	78,93	442,7333	10,66	6,274	73,33	1,02
Fora	May/June	25,9	6,94	28,5	594,4	4,24	3,88	39,894	0,8
	September/October	28,3	6,08	50,9	297,3	8,24	1,79	54,9	2,24
	December	22,9	6,91	93,9	719,3	5,28	4,89	39,894	1,84

Name of water points	Sampling period	T°C	pH	Conductivity (µS/cm)	Turbidity (NTU)	Calcium (mg/L)	Magnesium (mg/L)	Bicarbonate (mg/L)	Chloride (mg/L)
	Average values	25,7	6,64	57,76	537	5,92	3,52	44,896	1,62
Dam no. 2	May/June	30,6	7,68	161,6	800,4	17,28	5,424	68,686	9,3
	September/October	28,9	8,79	404	56,33	45,52	3,36	161,04	2,64
	December	24,9	6,8	419	113,8	43,2	3,23	159,64	1,53
	Average values	28,13	7,75	328,2	323,51	35,33	4,	129,78	4,49
Dam no. 3	May/June	28,7	7,71	294	224,4	21,92	4,32	116,876	29,7
	September/October	28,1	8,9	526	77,5	56,96	1,008	211,06	2,12
	December	25,1	6,98	397	61,82	48,52	1,005	197,85	2,07
	Average values	27,3	7,86	405,6	121,24	42,46	2,111	175,262	11,29
Nazinon to Ziou	May/June	26,7	7,68	32,4	692,8	14,8	0,384	18,422	0,5
	September/October	29,7	8,18	11,9	122,1	12,32	5,52	75,64	1,3
	December	28,6	7,9	12,88	118,2	10,34	4,91	55,22	0,65
	Average values	28,3	7,92	19,06	311,03	12,48	3,60	49,76	0,816
Wayen	May/June	28,1	8,4	66,1	870,3	6,72	3,792	35,014	1
	September/October	28,2	8,2	76,5	56,31	8,8	13,392	47,58	2,3
	December	25,8	8,3	83,2	255,9	10,22	2,82	28,45	1,98
	Average values	27,3	8,3	75,2	394,17	8,58	6,668	37,01	1,76
Loumbila dam	May/June	27,2	7,89	82,2	257,8	10	7,008	58,072	1,3
	September/October	34,6	7,95	56,1	458,1	4,72	3,072	25,62	0,78
	December	24,5	6,86	71,3	257,8	9,1	4,56	49,12	1,15

Name of water points	Sampling period	T°C	pH	Conductivity (µS/cm)	Turbidity (NTU)	Calcium (mg/L)	Magnesium (mg/L)	Bicarbonate (mg/L)	Chloride (mg/L)
	Average values	28,7	7,56	69,8	324,56	7,94	4,88	44,27	1,07
Bissiga	May/June	26,8	8,17	23,3	1100	5,44	4,272	25,254	0,3
	September/October	32,6	7,58	80,1	280	7,84	5,376	50,02	1,77
	December	25,4	6,89	103,1	112,7	6,45	3,27	24,4	1,3
	Average values	28,2	7,54	68,8	497,56	6,57	4,306	33,22	1,12
Bagré dam	May/June	27,2	7,4	79,6	327	8,64	7,536	41,968	1,9
	September/October	25,4	6,89	103,1	112,7	5,44	4,272	25,254	0,3
	December	25,8	6,97	62,72	272,7	9,55	8,56	45,64	2,5
	Average values	26,1	7,08	81,80	237,46	7,87	6,78	37,62	1,56
Goinré dam	May/June	26	7,3	62,1	1100	7,28	4,416	26,23	1,3
	September/October	28,1	8,51	55,7	301,2	28,48	9,408	32,94	1,18
	December	28,6	6,74	74,9	16,03	57,25	3,45	16,33	0,4
	Average values	27,5	7,51	64,23	472,41	31	5,758	25,16	0,96
Neboun	May/June	28,7	7,12	301	535,8	10,56	3,65	61,37	0,8
	September/October	23	7,3	67,6	166,2	8,65	1,65	51,26	1,5
	December	21,4	6,9	65,8	190,2	7,93	2,85	55,36	1,8
	Average values	24,3	7,10	144,8	297,4	9,04	2,71	55,99	1,366

Table IV: Analysis results for surface water sites (Continued)

Name of water points	Sampling period	Nitrate (mg/L)	Phosphate (mg/L)	Sulphate (mg/L)	Potassium (mg/L)	Sodium (mg/L)	Manganese (mg/L)	Copper (mg/L)	Zinc (mg/L)
Douna Barrage	May/June	0,1	0,2	0,2	2,7	< 0,5	< 0,05	< 0,05	< 0,05
	September /October	0,43	0,13	0,21	1,83	0,055	0,077	< 0,05	< 0,05
	December	0,38	0,18	0,19	2,23	< 0,5	< 0,068	< 0,05	< 0,05
	Average values	0,30	0,17	0,2	2,25				
Diarabakoko	May/June	2,3	0,3	1,2	3,35	1,28	0,05	< 0,05	< 0,05
	September /October	2,41	0,24	0,46	3,87	2,62	0,18	< 0,05	0,1
	December	2,37	0,36	0,56	3,75	2,38	0,15	< 0,05	< 0,05
	Average values	2,36	0,3	0,74	3,65	2,09	0,12		
Bougouriba	May/June	2	0,2	0,8	3,37	0,88	0,241	< 0,05	< 0,05
	September /October	1,27	0,19	0,3	2,86	3,725	0,059	< 0,05	< 0,05
	December	1,51	0,21	0,5	2,97	2,98	0,15	< 0,05	< 0,05
	Average values	1,59	0,2	0,53	3,06	2,52	0,15		
Mouhoun Boromo	May/June	1,3	0,2	0,9	3,12	< 0,5	0,2	< 0,05	< 0,05
	September /October	0,67	0,36	2,12	3,08	< 0,5	0,21	< 0,05	< 0,05
	December	1,11	0,28	2,09	3,05	< 0,5	0,26	< 0,05	< 0,05
	Average values	1,02	0,28	1,70	3,08		0,22		
Fora	May/June	3	0,2	1,2	2,89	< 0,5	0,103	< 0,05	< 0,05
	September /October	0,3	0,29	0,24	4,159	3,944	0,192	< 0,05	< 0,05

Name of water points	Sampling period	Nitrate (mg/L)	Phosphate (mg/L)	Sulphate (mg/L)	Potassium (mg/L)	Sodium (mg/L)	Manganese (mg/L)	Copper (mg/L)	Zinc (mg/L)
	December	4,11	0,36	1,43	3,94	< 0,5	0,15	< 0,05	< 0,05
	Average values	2,47	0,28	0,95	3,66		0,1483		
Dam no. 2	May/June	1,4	0,1	19,4	8,68	7,95	0,367	< 0,05	0,399
	September /October	0,1	0,23	0,6	20,35	6,744	0,285	< 0,05	< 0,05
	December	0,08	0,2	0,5	19,25	5,91	0,19	< 0,05	0,399
	Average values	0,526667	0,176	6,83	16,09	6,86	0,28		
Dam no. 3	May/June	1,5	0,2	45,4	15,1	27,2	0,14	< 0,05	< 0,05
	September /October	2,47	0,7	0,99	19,87	79,42	0,42	< 0,05	< 0,05
	December	2,3	0,21	1,04	18,15	65,42	0,34	< 0,05	< 0,05
	Average values	2,09	0,37	15,81	17,70	57,34	0,3		
Nazinon to Ziou	May/June	1,6	0,1	1,8	3,88	0,62	0,329	< 0,05	< 0,05
	September /October	1,26	0,37	0,61	4,132	5,237	0,202	< 0,05	< 0,05
	December	1,16	0,21	0,58	3,9	3,85	0,11	< 0,05	< 0,05
	Average values	1,34	0,22	0,99	3,97	3,23	0,21		
Wayen	May/June	2	0,1	3,8	4,6	1,04	0,152	< 0,05	< 0,05
	September /October	1,05	0,18	0,61	4,17	1,305	0,088	< 0,05	< 0,05
	December	2,1	0,21	0,81	3,6	0,04	0,16	< 0,05	< 0,05
	Average values	1,71	0,16	1,74	4,12	0,795	0,13		
Loumbila dam	May/June	1,7	0,1	2,4	6,95	1,74	0,123	< 0,05	< 0,05
	September /October	0,69	0,14	0,4	2,75	0,17	0,05	< 0,05	< 0,05

Name of water points	Sampling period	Nitrate (mg/L)	Phosphate (mg/L)	Sulphate (mg/L)	Potassium (mg/L)	Sodium (mg/L)	Manganese (mg/L)	Copper (mg/L)	Zinc (mg/L)
	December	0,75	0,11	0,6	4,87	1,24	0,11	< 0,05	< 0,05
	Average values	1,04	0,11	1,13	4,85	1,05	0,09		
Bissiga	May/June	2,6	0,1	1,7	2,7	< 0,5	0,3	< 0,05	< 0,05
	September /October	0,79	0,26	0,53	4,83	1,291	0,073	< 0,05	< 0,05
	December	0,6	0,11	0,7	3,78	0,5	0,3	0,05	< 0,05
	Average values	1,33	0,15	0,97	3,77		0,22		
Bagré dam	May/June	2,1	0,1	5,4	4,54	2,87	0,129	< 0,05	< 0,05
	September /October	2,6	0,1	1,7	2,7	< 0,5	0,3	< 0,05	< 0,05
	December	2,9	0,1	4,5	5,14	3,8	0,22	< 0,05	< 0,05
	Average values	2,53	0,1	3,86	4,126		0,216		
Goinré dam	May/June	9,1	0,3	2,6	5,5	0,64	0,197	< 0,05	< 0,05
	September/October	2,18	0,17	1,12	2,95	1,493	0,077	< 0,05	< 0,05
	December	1,2	0,2	1,6	1,5	0,67	< 0,05	< 0,05	< 0,05
	Average values	4,16	0,22	1,77	3,31	0,93			
Neboun	May/June	3,2	0,1	0,7	2,8	1,1	< 0,05	< 0,05	< 0,05
	September /October	0,9	0,1	0,1	2	4,7	0,057	< 0,05	< 0,05
	December	1,1	0,1	0,5	3,1	3,7	0,05	< 0,05	< 0,05
	Average values	1,73	0,1	0,43	2,63	3,16			

II.3. Quality control of test results

Ionic balances were determined for all the data in order to assess their acceptability. The ionic balance is the ratio of the sum of cations minus the sum of anions to the sum of cations and anions. It is expressed as follows:

$$BI(\%) = \frac{\Sigma cations - \Sigma anions}{\Sigma cations + \Sigma anions} X\,100$$

With :

Σcations: sum of cations (Ca^{2+}; Mg^{2+}; Na^{+}; K^{+}; Fe^{2+} and Mn^{2+}) in meq/L

Σanions: sum of anions (HCO_3^{-} ; SO_4^{2-} ; NO_3^{-} and Cl^{-}) in meq/L

As a general rule, the results of analyses are considered as follows:

-1%<BI<1% : Excellent reliability ;

-5%<BI<5% : Acceptable reliability ;

-10%<BI<10%: Poor reliability ;

-BI < -10% or BI>10%: Poor reliability.

If the value of the balance is of the order of 5%, the analyses are deemed acceptable.

As part of the 2023 monitoring, 80% of analyses are of acceptable quality.

III. PROCESSING AND INTERPRETATION OF RESULTS

III.1. Hydrochemical characteristics of water from surfaces

III.1.1. Traditional physico-chemical variables (temperature, pH, conductivity, turbidity, alkalinity)

III.1.1.1. Temperature

Water temperature plays an important role in the solubility of salts and oxygen, which are necessary for the balance of aquatic life. The value of the water temperature is generally influenced by the ambient temperature, but also by any discharges. The temperatures recorded ranged from 21.4°C (at Neboun in December) to 34.6°C (at Loumbila dam in October **(Figure 2)**. Temperatures fell in December due to the cooler weather at this time of year.

III.1.1.2. pH

The pH is a measure of the acidity of water, i.e. the concentration of H ions^{+} . In practice, the pH scale ranges from 0 (very acidic) to 14 (very alkaline). The median value of 7 corresponds to a neutral solution at 25°C.The pH of natural water can vary from 4 to 10 depending on the acidic or basic nature of the ground through which it flows. Low pH levels (acidic water) increase the risk of metals being present in the water, while high pH levels increase concentrations of ammonia, which is toxic to fish. For all three campaigns, the pH data show that the surface water is slightly neutral to alkaline, with pH values between 6.08 and 8.9 (**Figure 3**).

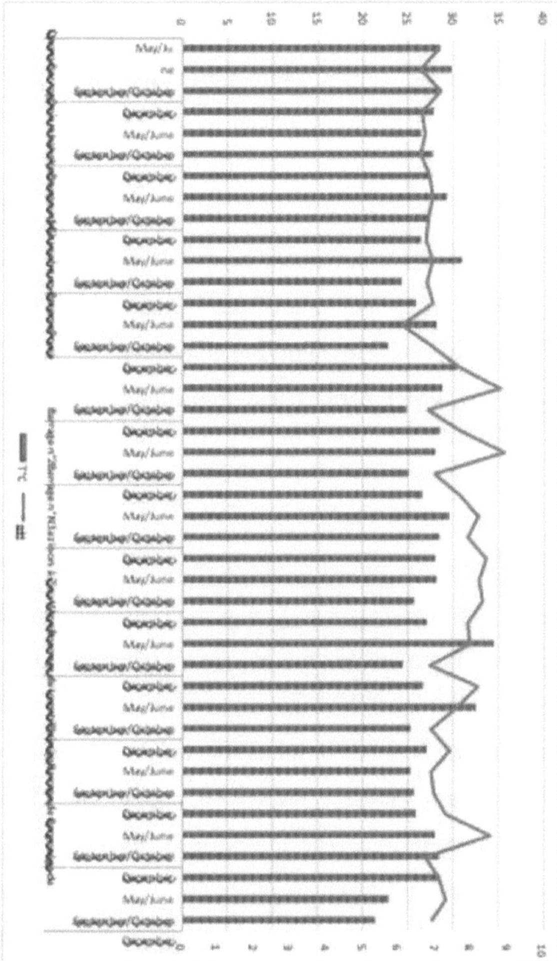

Figure 4: Spatiotemporal variations in temperature and pH of the surface waters studied

III.1.1.3. Conductivity

Electrical conductivity is a numerical expression of the capacity of a solution to conduct electric current. It is closely linked to the concentration of dissolved minerals in the water. Changes in the conductivity of a body of water can reveal signs of domestic or agricultural pollution. The conductivity values recorded vary from 11.9 µS/cm (at Ziou in September) to 526 µS/cm (at Barrage N°3 in September) **(Figure 3)**. The amplitude of the variation is greater during low-water periods. This reflects the high mineralisation of the water due to evaporation during low-water periods and low mineralisation due to run-off during the rainy season.

III.1.1.4. Turbidity

It can be assessed by the transparency of the water. Turbidity is a limiting factor for photosynthesis and therefore production, except when it is due to production (primary phytoplanktonic) itself. It is due to the presence of suspended particles in the water, which may be organic or mineral in origin. Water turbidity depends on the nature of the terrain it crosses, the season, rainfall, the water flow regime and the nature of the discharge. The lowest turbidity value was recorded in December at the Goinré dam in Ouahigouya (16.03 NTU), while the highest value was recorded at Bissiga in the centre-north and at the Goinré dam in June (1,100 NTU) **(Figure 3)**. The highest values were also recorded during the May/June campaign on the Mouhoun river at Boromo (1058 NTU), on the Wayen bridge (870 NTU) and at Barrage N°2 in Ouagadougou (800.4 NTU) **(Figure 3)**. The highly turbid nature of the surface water in June is due to the presence of a large quantity of suspended matter from rainwater run-off.

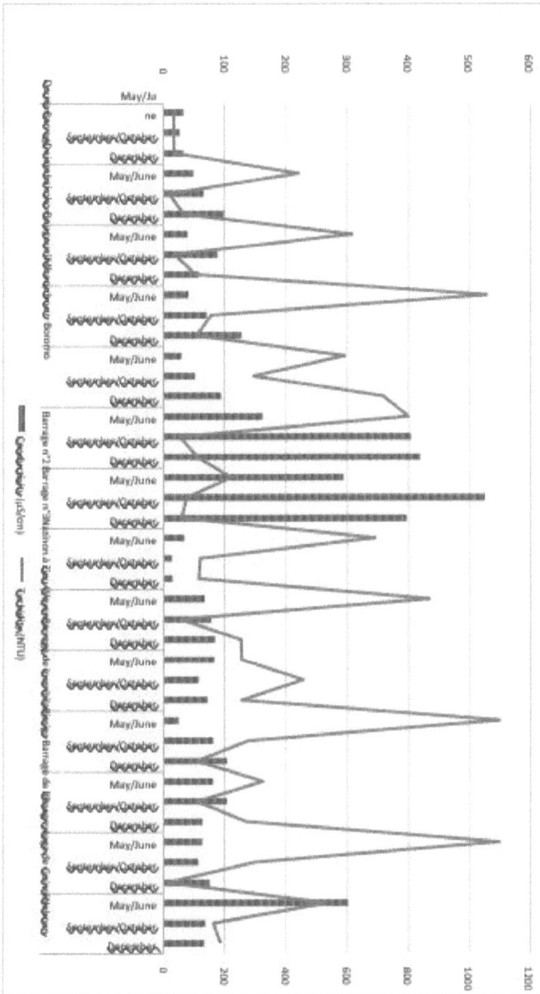

Figure 5: Spatiotemporal variation in conductivity and turbidity in the surface waters studied

III.1.1.5. Alkalinity

Figure 4 shows the spatial variations in the results obtained from analyses of calcium (TCa) and magnesium (TMg) titres and bicarbonates (HCO3-) in the surface waters studied. These results show a correlation to a greater or lesser extent between HCO3- values and aPTT and MGT values. HCO3 levels in surface water comply with the 2017 WHO guide values of 500 mg/L. Similarly, the calcium and magnesium values comply with the WHO guide values, which are 75 mg/L and 50 mg/L respectively, and the national guide values.

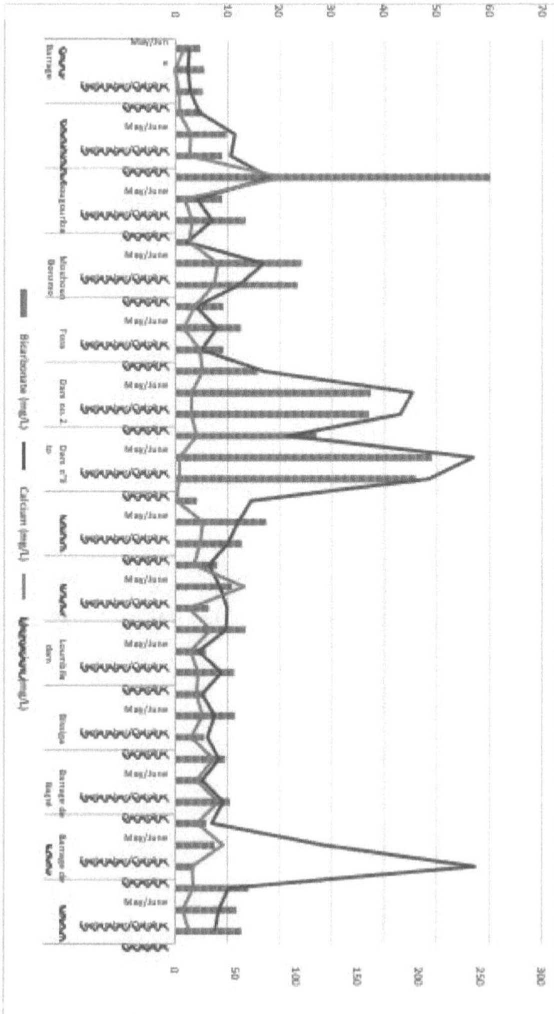

Figure 6: Spatiotemporal variation in bicarbonate, calcium and magnesium ion content in the surface waters studied.

III.1.2. Major ion content

The mineralisation of most water is dominated by eight ions, commonly known as major ions. These are cations: calcium, magnesium, sodium and potassium, and anions: chloride, sulphate, nitrate and bicarbonate.

Ions naturally vary widely in surface and groundwater due to geological, climatic and geographical conditions.

- Knowledge of sodium (Na^+) concentrations is of particular interest when water is used for drinking or irrigation.
- Concentrations of potassium (K^+) are generally very low, except in brine waters or hot springs, or when they are polluted by inorganic fertilisers or industrial effluents.
- Chloride (Cl^-) levels are generally low in surface waters, but can be increased due to the deposition of oceanic aerosols, industrial and wastewater effluents and leaching.
- Industrial discharges and atmospheric precipitation can add significantly to the natural levels of sulphates (SO_4^{2-}) in surface water. High concentrations (> 400 mg. l^{-1}) can make water unpalatable. Some bacteria use sulphate as a source of oxygen and convert it to hydrogen sulphide (H_2S) under anaerobic conditions. Hydrogen sulphide is a colourless gas with an unpleasant smell of rotten eggs, which is often indicative of the presence of anaerobic conditions. In the aquatic environment, the main nutrients are nitrogen and phosphorus compounds. Nitrogen compounds come from the drainage of soil, plant and animal debris. Phosphorus compounds come mainly from the decomposition of organic matter.

III.1.2.1. Chlorides

The presence of chlorides in water can result from marine intrusion, the dissolution of evaporites or the alteration of minerals in the surrounding rocks. It can also come from anthropogenic sources. The chloride levels measured were more or less identical for the three campaigns. However, high chloride concentrations of the order of 9.3 mg/l and 29.7 mg/l were noted respectively at dams N°2 and N°3 in Ouagadougou in the month of June **(Figure 7).** This could be explained by pollutant discharges from the city, which are the source of chloride inputs to the dams at the start of the rainy season.

III.1.2.2. Nitrates

Nitrates are forms of nitrogen that occur naturally in the environment. Nitrate is essential for plant growth and is present in all vegetables and grains. It is commonly used in fertilisers.

The most common sources of nitrate in water are :

- Chemical fertilisers used to improve crop growth;

- Animal waste from barns and manure storage sites;

- Waste of human origin from septic fields, septic tanks or leaky holding tanks;

Nitrates are found in surface water at higher levels in May. /June compared with the September/October and December periods. It varied between 0.1 and 9.1 mg/L over the three seasons. The increase in nitrate levels in May could be explained by rainwater runoff at the beginning of the season. The high value of 9.1 mg/L (**Figure 8)** was observed at the Goinré dam in May/June. This high level is thought to be due to the use of fertilisers for agricultural practices and irrigated farming on a developed area around the dam.

III.1.2.3. Sulphates

Sulphate levels were relatively low in all three years. However, high values were recorded at dams N°2 and N°3 with values of 19.4mg/L and 45.4mg/L respectively in June **(Figure 9)**. This could be explained by pollutant discharges from the town from market garden crops, which are the source of sulphate inputs to the dams at the start of the season.

III.1.2.4. Phosphates

Phosphate levels in surface water were low over the three surveys, ranging from 0.1 mg/L in May to 0.7 mg/L in September **(Figure 10)**. However, the highest value was observed at Ouagadougou dam N°3 with 0.7 mg/L.

III.1.2.5. Sodium

Sodium is a constant element in water. However, concentrations can be extremely variable, ranging from a few tens of milligrams to 500 mg/L and even higher. Apart from the leaching of geological formations containing sodium chloride, this salt can come from the decomposition of mineral salts such as sodium and aluminium silicates. Sodium levels were more or less the same during the three surveys, but very high values of around 27.2 mg/L in June, 79.42 mg/L in September and 65.42 mg/L in December were observed at Ouagadougou dam N°3 **(Table IV)**. The maximum is observed during high-water periods, which can be explained by the contribution of pollutant discharges from the city via rainwater runoff.

III.1.2.6. Potassium

Potassium is less abundant than sodium. It is rarely present at levels above 20 mg/L in natural waters.It is frequently present in rocks and soils. It can also come from weathering and erosion of potassium-containing minerals such as feldspar, and leaching from soils containing fertilisers. Potassium concentrations varied between 1.5 mg/L and 20.4 mg/L **(Table IV)** during the three campaigns. The highest concentrations were observed at dams N°2 and N°3 in Ouagadougou, with values of :

- 8.68 mg/L in June and 20.35 mg/L in September and 19.25 in December for dam N°2

-15.1mg/L in June and 19.87 mg/L in September and 18.15 in December dam N°3(Table IV).

These high values can be explained by the contribution of pollutant discharges from the city of Ouagadougou via rainwater run-off.

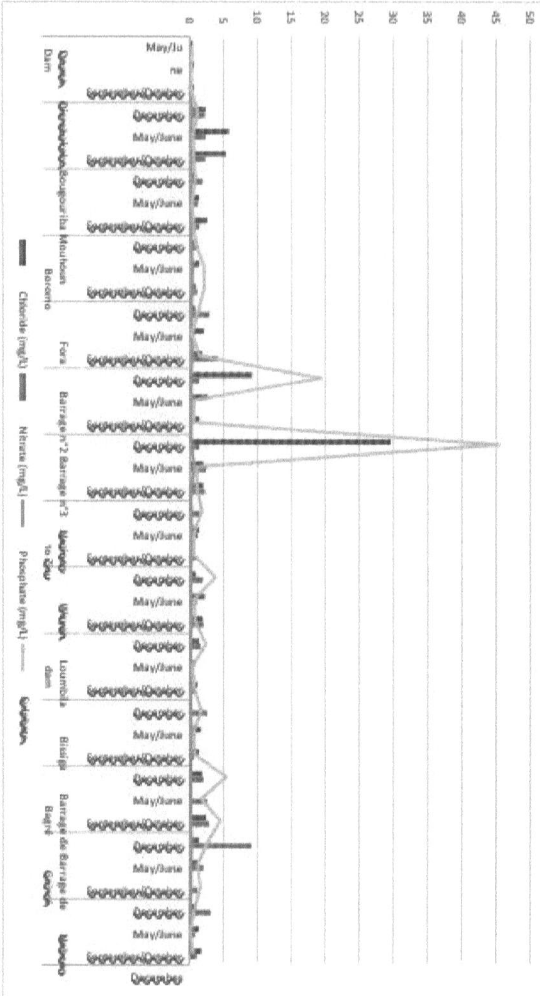

Figure 11: Spatial-periodic variation in chloride, nitrate, phosphate and sulphate ions in the surface waters studied

III.1.3. Heavy metal content in surface water

Monitoring concentrations of heavy metals (density > 5 g/cm3) is particularly important given their toxicity and capacity to bioaccumulate along food chains. Unlike organic pollutants, metals cannot be biologically or chemically degraded. Traces of metals are always present in freshwater from the weathering of rocks and soil. Other sources of heavy metals include wastewater discharges and mining. The ability of a water body to support aquatic life is dependent on the availability of trace elements, notably manganese (Mn), and zinc (Zn), but at high concentrations such elements can be toxic to humans and other forms of life. Some elements, such as mercury (Hg), can accumulate in high concentrations in the tissues of aquatic organisms.The presence of heavy metals in water is characteristic of certain types of pollution due to industrial, mining, tannery or dyeing discharges. Concentrations of copper, zinc and manganese are practically zero, as most values are below the instrument's detection limit of 0.05mg/L **(table IV).**

For the surface waters studied, there was a total absence of copper and zinc over the three campaigns. A maximum content of 0.41 mg/L was observed in June at Ouagadougou dam N°3 **(Table IV)**. This value can be explained by the phenomenon of evaporation, which reduces the quantity of water in the structure while increasing the content of the various chemical elements. The origin of manganese in surface water could be the result of rainwater transport of domestic and industrial waste.

III.2. Hydrochemical characteristics of groundwater

III.2.1 Traditional physico-chemical variables (temperature, pH, conductivity, turbidity, alkalinity)

III.2.1.1. Temperature

Groundwater temperatures varied between 25.8°C and 32.4°C over the three campaigns **(Figure 12)**. The values were more or less the same over all three periods.

III.2.1.2. pH

The values observed show that most groundwater pH values are close to neutral. The pH ranges from 5.16 to 8.5 **(Figure 13)**. The majority of pH values do not exceed the WHO standard for groundwater ($5.5 \leq pH \leq 8$). However, basic pH values were observed at the Louda and Bassinko piezometers, with values of 8.1 and 8.5 respectively. These high pH values could be explained by the nature of the water table.

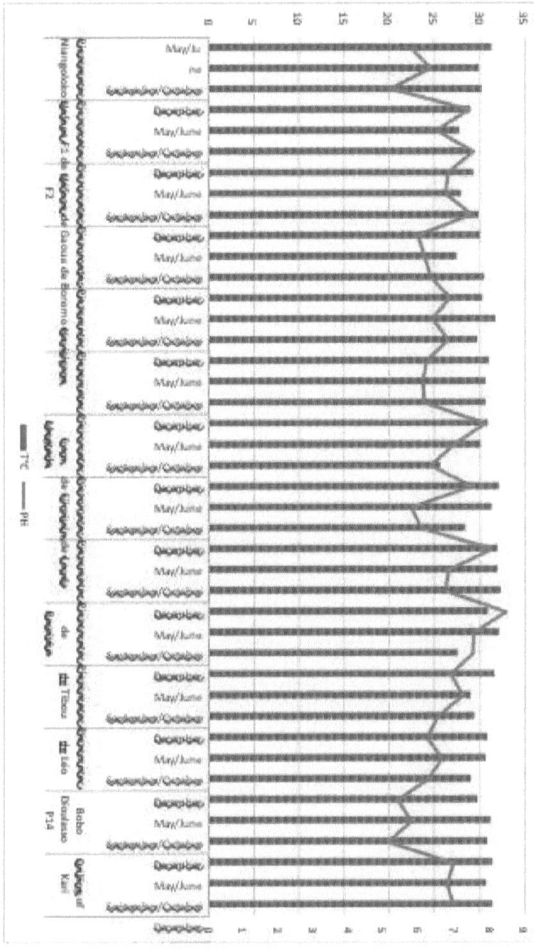

Figure 14: Spatiotemporal variations in temperature and pH of the groundwater studied

III.2.1.3. Conductivity

Electrical conductivity (EC) is a measure of the ability of an aqueous solution to conduct an electrical current, which depends on the concentration of ions in solution. It is sensitive to variations in dissolved matter, mainly mineral salts, and can be useful for characterising water quality.

Conductivity reflects the total mineralisation of the water (BAWAR, 2022).

- 50 to 400 µS/cm: excellent qualities,
- 400 to 750 µS/cm: good quality,
- 750 to 1500 µS/cm: poor quality but usable water,
- From 1500 µS/cm and above: excessive mineralisation.

Groundwater conductivities ranged from 56.2µS/cm to 524µS/cm. **(Figure 7)** The lowest value recorded was at the Bobo piezometer, while the highest was at the Boromo piezometer. Overall, these values are in line with natural conductivity guide values. The majority of groundwater therefore has a conductivity of between 50 and 400 µS/cm. This would indicate good quality water.

III.2.1.4. Turbidity

Turbidity values varied between 0.2 and 402 NTU over the three surveys **(Figure 7)**. There was an increase in turbidity during the dry season, which can be explained by the drop in the water table. The highest value was recorded at Bobo piezometer P14, which is over 80 metres deep. This extreme value may be due to the nature of the water table with a wave of infiltration of water from the underground watershed.

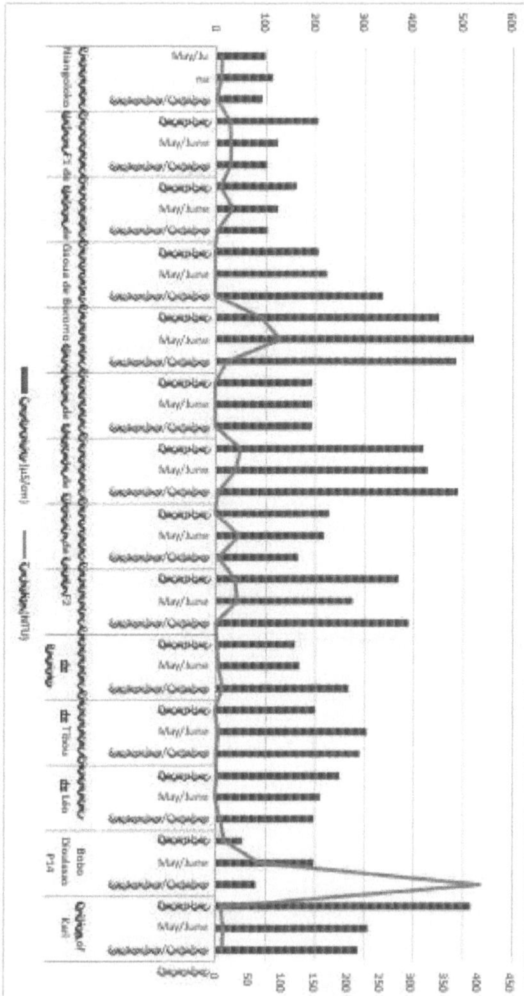

Figure 15: Spatiotemporal variations in turbidity and conductivity of the groundwater studied

III.2.1.5. Alkalinity

Alkalinity is a measure of water's ability to neutralise acids. The neutralising power of water is mainly attributed to the presence of bicarbonate ions in water. Hardness refers to the calcium and magnesium content of water. Calcium ($Ca2+$) is an essential element for living organisms and is incorporated into the bones and shells of invertebrates. Calcium can precipitate out of water when there is a drop in carbon dioxide levels, for example due to photosynthetic activity.Magnesium ($Mg2+$) is present in natural water and contributes to hardness, along with calcium. Natural levels are rarely significantly influenced by industrial effluents.

Figure 8 shows a correlation between the calcium, magnesium and bicarbonate values of the groundwater studied. The high values of these ions indicate a more or less good quality of the groundwater. Overall, HCO3 levels in the groundwater studied were in line with the WHO 2017 guide, which is 500 mg/L **(Figure 8).**

Similarly, most calcium and magnesium levels comply with the WHO guide values of 75 mg/L and 50 mg/L respectively. This would indicate that total hardness is compliant **(Figure 8).** However, higher or lower values of these calcium ions were observed in the water from the Kari borehole (mean value of 50.71 mg/L) and the Mogtedo piezometer (mean value of 84.14 mg/L). This may be due to the nature of the aquifer crossed. The instability of the balance of these waters could be due to mining activity in these areas.

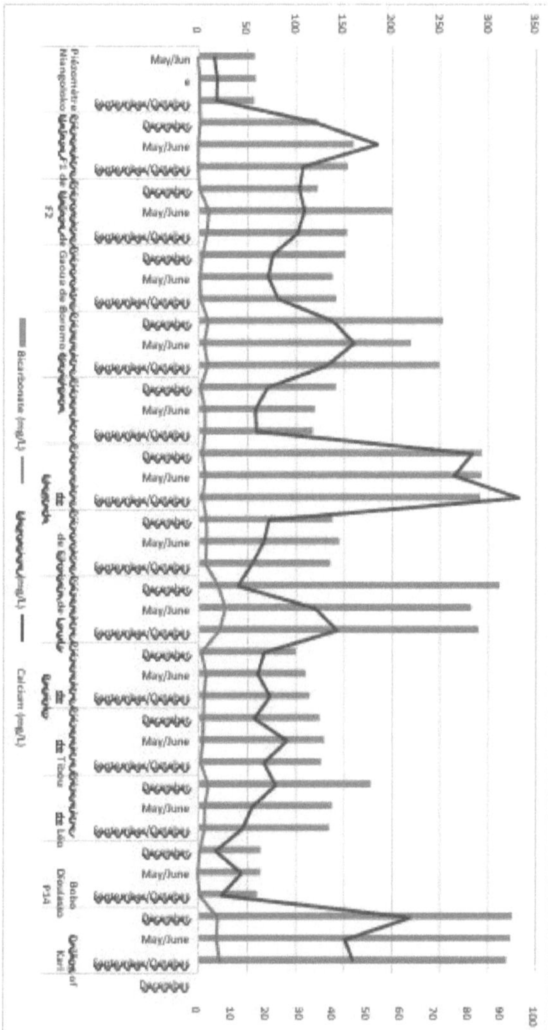

Figure 16: Spatio-period variation in bicarbonate, magnesium and calcium ion content in the groundwater studied

III.2.2. Levels of other major ions in groundwater

III.2.2. 1.chlorides

Chloride concentrations ranged from 0.1 to 85.2mg/L over the three campaigns. The maximum value was observed at the Silmissin piezometer, with a level of 85.2mg/L **(Figure 9)**. The chloride values recorded comply with the drinking water standard of 250mg/L.

III.2.2.2. Nitrates

Nitrates are relatively low in groundwater, with levels varying between 0.1 and 20.5 mg/L over the three surveys **(Figure 9).** All the groundwater meets the drinking water standard in force in Burkina Faso for nitrates, which is 50 mg/L.

III.2.2.3. Phosphates

Phosphate concentrations were more or less the same over the three seasons, with a slight increase in September. This could be explained by the leaching of chemical fertilisers used in agriculture. Levels varied between 0.23 and 2.3mg/L over the three campaigns **(Figure 9)**. It should be noted that the drinking water standard does not set any standards for phosphates; the imperative guide value would be 0.4 mg/L.

III.2.2.4. Sulphates

Sulphate concentrations are relatively identical throughout the seasons, with values ranging from 0.1 to 16.95 mg/L **(Figure 9).** Sulphate levels comply with the standard of 250mg/L.

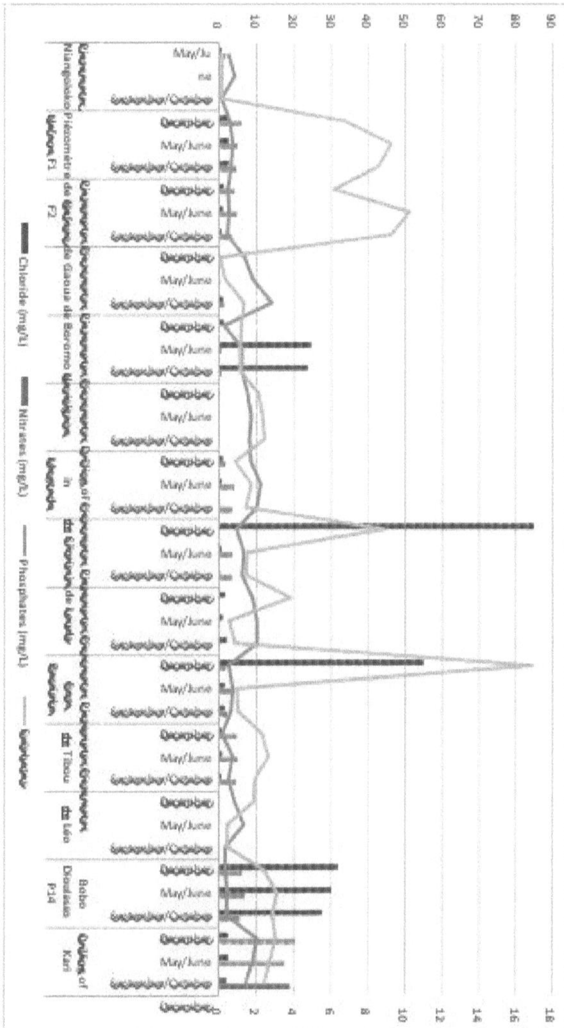

Figure 17: Spatio-temporal variations in chloride, nitrate, phosphate and sulphate levels in the groundwater studied

III.2.2.5. Potassium

Potassium concentrations were relatively identical over the three surveys, varying from 0.5 to 14.5 mg/L over the three surveys. The two highest values were observed in the Boromo (12.5 mg/L) and Léo (14.5 mg/L) piezometers **(Table II).** However, it should be noted that there is no drinking water standard for potassium; the imperative guide value would be 200 mg/L.

III.2.2.6. Sodium

Sodium concentrations varied between 0.5 and 29.31 mg/L over the three campaigns. (Table II).The maximum value was observed at the Tibou piezometer **(Table II).** Sodium concentrations comply with the drinking water standard of 200 mg/L.

III.2.3. Heavy metal content in groundwater

The concentrations of copper and zinc in the groundwater were practically zero during both periods, as they were below the instrument's detection limit of 0.05mg/L **(Table II).**

All these values comply with the standards, which are 1 mg/L for copper and 3 mg/L for zinc.Manganese concentrations are mostly below the standard, which is 0.5 mg/L. However, values above the standard were observed in the Niangoloko (0.89 mg/L) and Louda (1.37 mg/L) piezometers **(Table II).**
The high levels of manganese are due to the nature of the rocks through which it passes.

IV. CONSTRAINTS

The major difficulties that have hindered the implementation of the activities include the following:

- Repetitive breakdowns on analysis equipment;

- Lack of funds for maintenance;

- Lack of expertise in the use of certain laboratory equipment by staff due to the failure to train new staff in the use of this equipment, especially heavy equipment;

- Mobility of laboratory staff due to lack of motivation;

- Depletion of stocks of reagents and consumables;

- Insufficient logistical resources (vehicles) to carry out field missions;

- The small number of laboratory staff (four) means that it is not possible to carry out all the laboratory's activities properly (cleaning the sample glassware, handling the equipment at the analysis stations, entering and processing data, etc.);

- The virtual non-existence of maintenance services on the electricity and water supply network to ensure the functionality of the DGRE laboratory building;

- Insecurity in certain sample areas.

V. RECOMMENDATIONS AND OUTLOOK

With a view to better carrying out the department's activities, which can be ummed up as follows mainly for laboratory services, it is necessary :

- Speed up the process of releasing budgets and streamline procurement procedures, so that the laboratory's equipment and consumables are available in time to provide these services;
- Continuing efforts to make the DGRE laboratory operational by supplying an inductively coupled plasma mass spectrometer (ICP-MS), a new ion chromatograph or High Performance Liquid Chromatography (HPLC), an ultrapure water distiller, accessories for the Gas Chromatograph and groundwater sampling pumps.
- Recruit and train laboratory staff on an ongoing basis in the handling and preventive maintenance of equipment, and motivate staff to minimise their mobility;
- Drawing up and allocating a budget for the maintenance and repair of laboratory equipment;
- Draw up maintenance and repair plans with manufacturers or their representatives;
- Draw up and implement the laboratory's organisational and functional texts;
- Recruiting expertise to support the design and implementation of a quality management according to the 17025 standard;
- Work to strengthen collaboration between the various structures involved in the analysis of water at national and international level.
- Resolving problems on the electricity and water supply networks to ensure continuity of service functionality of the laboratory building.

CONCLUSION

This work focused on the quality of surface and groundwater in the water quality monitoring network of the Directorate-General for Water Resources (DGRE). It consisted of carrying out three water sampling campaigns in the network, followed by in situ and laboratory analysis, except at sites located in insecure areas of the country. For all the waters studied, it appears that for surface waters, certain sites (dams N°2 and N°3 in Ouagadougou and the Goinré dam in Ouahigouya) are subject to pollution of anthropic origin (pollution by nitrates, sulphates, chlorides, phosphates, sodium, potassium and heavy metals). Groundwater quality is satisfactory, as most of the parameters analysed comply with the drinking water standards in force in Burkina Faso. However, most of the groundwater studied had high turbidity levels that exceeded the norm. The Boromo, Mogtedo, Nafona F1, Niangoloko and Louda piezometers have turbidities in excess of 5 NTU ;The Niangoloko and Louda piezometers have manganese levels above the standard.In all, less than 50% of the groundwater analysed met the drinking water standards in force in Burkina Faso. Given the importance of water quality knowledge, the DGRE laboratory needs to be provided with sufficient material, human (qualified and well-trained) and financial resources to ensure that the country's water quality situation is properly assessed and that the quality network is constantly monitored. It is also important to look at the possibility of monitoring the new optimised quality network and the pilot networks on mining sites and industrial discharges launched in 2021, in order to determine the overall quality of Burkina Faso's water resources.

REFERENCES BIBLIOGRAPHY

- DGRE (Direction Générale des Ressources en Eau), 2020, Rapport sur l'état des lieux de la qualité des eaux brutes du bassin du Mouhoun : rapport d'inventaire partiel, 178p ;
- Joint decree N°00019/MAHRH/MS defining water potability standards in force since 05 April 2005;
- Burkinabe discharge standards taken from DECREE N°2015- 1205 IPRES-TRANS/PMI MERH IMEF/MARHASAIMS/MRA/MICAI MME/MIDT/MATD on wastewater discharge standards and conditions;
- WHO STANDARDS (2017), some physico-chemical parameters of water ;
- BAWAR B., 2022; Stratégie de Management communautaire des ressources en eau, specific document on water resource management, published by EUE, 2022. ISBN ; 978- 613-9-50139-7, 61p.

APPENDICES

APPENDIX 1: TABLE I: SOME OF THE EQUIPMENT AND ANALYSIS METHODS USED

Parameters analysed	Units	Methods	Standards	Equipment used	Limits of detection
Temperature	°C	Electrochemical	NF EN 27888	3210 conductivity meter	0.0 à 60.0
pH		Electrochemical	NF T 90-008	Hanna HI 98128	-2.00 à 16.00
Conductivity	µS/cm	Electrochemical	NF EN 27888	3210 conductivity meter	0.0 à 3999
Turbidity	NTU	Nephelometric	NF EN ISO 7027	355 IR Turbidimeter	0.02
Chloride	mg/l	Chromatography/ spectrophotometry	HACH method 8048	Metrohm/DR3900 IC Ion Chromatograph	0.02
Nitrate	mg/l	Chromatography/ spectrophotometry	HACH method 8114	Metrohm/ DR3900 IC Ion Chromatograph	0.3
Phosphate	mg/l	Chromatography/ spectrophotometry	HACH method 8008	Metrohm/ DR3900 IC Ion Chromatograph	0.02
Sulphate	mg/l	Chromatography/ spectrophotometry	HACH method 8051	Metrohm/ DR3900 IC Ion Chromatograph	2
Sodium	mg/l	Absorption spectrometry atomic	FD T90-112	SAA Perkin Elmer Pin AAcle 900T	0.5
Potassium	mg/l	Absorption spectrometry atomic	FD T90-112	SAA Perkin Elmer Pin AAcle 900T	0.5
Copper (Cu)	mg/l	Absorption spectrometry atomic	FD T90-112	SAA Perkin Elmer Pin AAcle 900T	0.05
Manganese (Mn)	mg/L	Absorption spectrometry atomic	FD T90-112	SAA Perkin Elmer Pin AAcle 900T	0.05
Zinc (Zn)	mg/l	Absorption spectrometry atomic	FD T90-112	SAA Perkin Elmer Pin AAcle 900T	0.05

Appendix 2: Some illustrative photos, DGRE 2023: equipment, analyses, field sampling, in-situ measurements

Figure 10:

Turbidity meterfigure 11: Conductivity meter

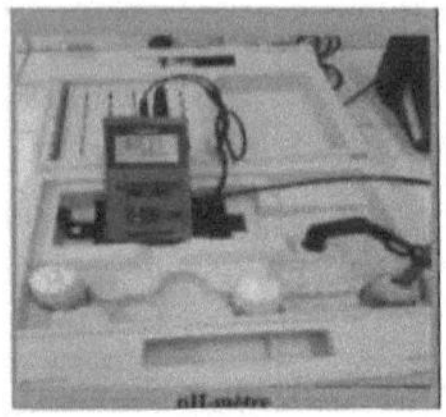

pH-mètre

Garmin 62x GPS

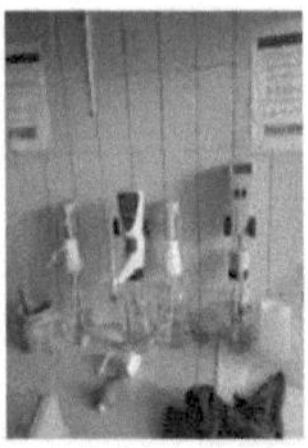

Burettes for measuring alkalinity

IC Metrohm ion chromatograph

SAA Atomic Absorption pectrometer Perkin Elmer Pin AAcle 900T

In situ analysis

On the [illegible] Piezometer

Ready for spectrometer analysis atomic absorption

Profile view of the [illegible] dam in December

Sampling in the Mouhoun river at Boromo

Median view of the Bagré dam in December.

MIX
Papier aus verantwortungsvollen Quellen
Paper from responsible sources
FSC® C105338

Printed by Books on Demand GmbH, Norderstedt / Germany